W9-BVI-214

7

Kansas City, MO Public Library
00001841155525

Building ON A DREAM

THE EIFFEL TOWER

Russell Roberts

PURPLE TOAD
PUBLISHING

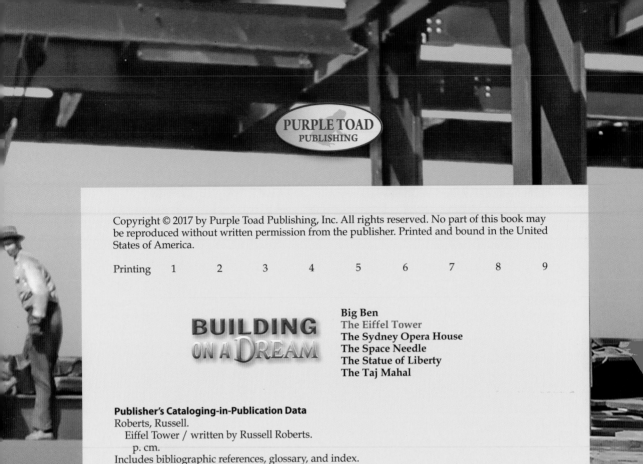

PURPLE TOAD
PUBLISHING

Copyright © 2017 by Purple Toad Publishing, Inc. All rights reserved. No part of this book may be reproduced without written permission from the publisher. Printed and bound in the United States of America.

Printing   1   2   3   4   5   6   7   8   9

BUILDING ON A DREAM

Big Ben
The Eiffel Tower
The Sydney Opera House
The Space Needle
The Statue of Liberty
The Taj Mahal

Publisher's Cataloging-in-Publication Data
Roberts, Russell.
  Eiffel Tower / written by Russell Roberts.
    p. cm.
Includes bibliographic references, glossary, and index.
ISBN 9781624692031
1. Tour Eiffel (Paris, France)—Juvenile literature. 2. Eiffel Tower (Paris, France). 3. Architecture—Vocational guidance—Juvenile literature. I. Series: Building on a Dream.
 NA2555 2017
 507.8
                                        Library of Congress Control Number: 2016937176

eBook ISBN: 9781624692048

ABOUT THE AUTHOR: Russell Roberts has researched, written, and published numerous books for children and adults. Among his books for adults are *Down the Jersey Shore*, *Historical Photos of New Jersey*, and *Ten Days to A Sharper Memory*. He has written over 50 nonfiction books for children. Roberts often speaks on the subjects of his books to various groups and organizations. He lives in New Jersey.

CONTENTS

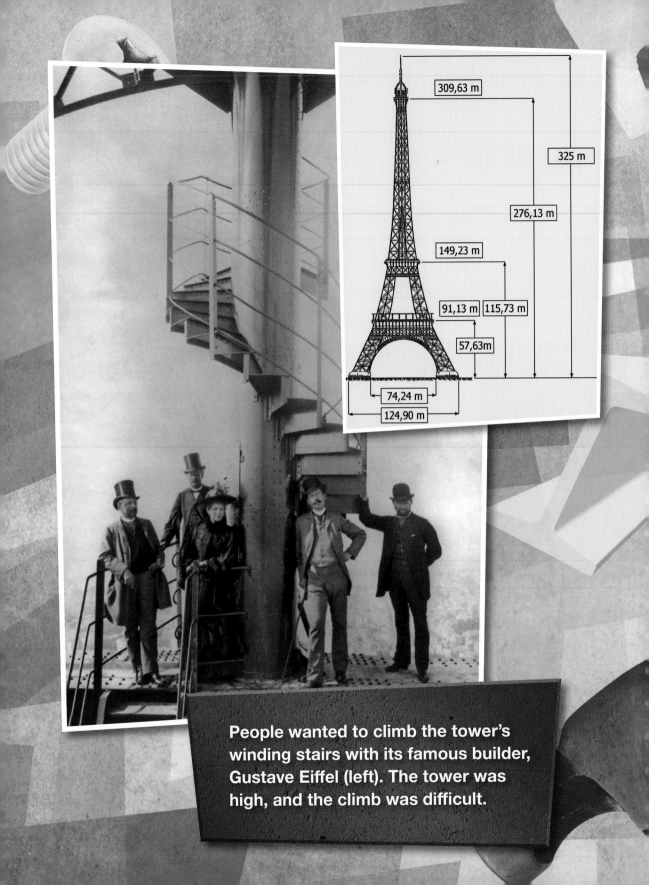

309,63 m

325 m

276,13 m

149,23 m

91,13 m 115,73 m

57,63m

74,24 m

124,90 m

People wanted to climb the tower's winding stairs with its famous builder, Gustave Eiffel (left). The tower was high, and the climb was difficult.

The Terrifying Climb

February 24, 1889, was a cold day in Paris, France. At 8:00 a.m., the temperature was two degrees below zero. Snow fell from a gloomy sky.

None of this mattered to journalist Robert C. Henri. Today he was going to meet the famous French builder Gustave Eiffel. And today he would become the first reporter to climb to the top of the 1,000-foot metal tower that Eiffel was building for the Industrial Exhibition.

Henri's horse-drawn carriage arrived to pick him up, the horses' breath rising from their nostrils in white clouds of steam. When Henri's carriage reached the tower, Eiffel was waiting for him, along with 15 other men and women.

At 2:30 p.m. the group began their single-file climb. Eiffel advised the others to walk as he did. He grasped the railing with his right arm and swung his body from side to side. The first part of the climb was gradual, and the group ambled up.

When they reached the first platform, Henri looked down. From 190 feet above the ground, the people below looked like tiny black spots. "Only the rippling Seine [the nearby river] seemed still alive," he wrote.[1]

The next part of the climb was a spiral staircase. Five people stayed behind on the first platform, and the rest continued. At 3:45 p.m., the climbers reached the tower's second platform, which was nearly 380 feet high. From there in the fading winter light, Henri could see the roofs of many Paris buildings. He looked through a small opening in the floor down to the ground. "A shiver ran down my spine

The winding tower steps

at the thought of a possible fall from this height. It grew suddenly colder."[2]

Once again the group set out. By then it was 4:10 p.m. The wind was howling like a wounded animal, grabbing and tearing at Henri's clothes. Then the sky opened and hail rained down, making it difficult to see. Ice coated the iron railing.

At 5:00 p.m. they reached a temporary platform, but since the light was fading, the group decided to go immediately to the top. They found that the next iron staircase was not attached at the bottom—just the top. It swayed back and forth in the terrific wind.

This was too much for several members of the group. Only Henri, Eiffel, and two others decided to continue to the top. Then the staircase ended. They had to climb the rest of the way on ladders tied together with heavy rope.

"Look not to the right nor to the left!" Henri told himself. "Keep your eyes only on the rung of the ladder on which you are about to place your foot!"[3]

After climbing three ladders, the four reached the final platform, nearly 900 feet above the ground. Henri heard the faint sounds of church bells ringing, but he could see nothing in the darkness below. The platform bounced and rocked in the wind. He felt he was on a ship

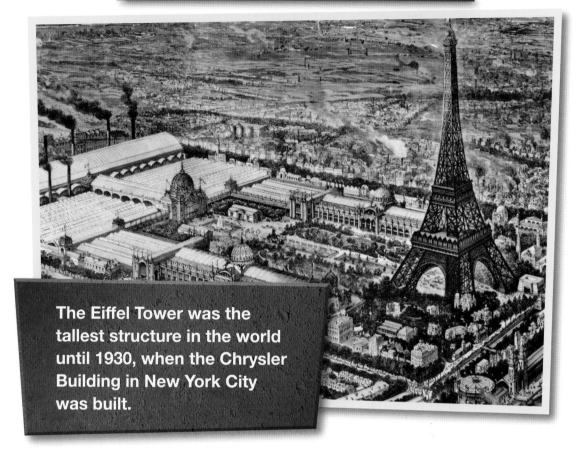

The Eiffel Tower was the tallest structure in the world until 1930, when the Chrysler Building in New York City was built.

rolling on the waves. He cautiously approached the edge of the platform to look down.

Suddenly Henri felt himself toppling forward. He frantically thrust his hand out and was relieved to grab a rope that was hanging near the side. Then the rope began to slide down, and Henri went with it!

"In a frenzy of terror . . . I felt myself falling!" he wrote.[4]

Henri felt others next to him. A moment later, now pulled safely back from the edge, he heard Eiffel say: "You should never touch a rope—that one is attached only to a pulley. Had you leaned more heavily on it the consequences would not have been pleasant."[5]

Five weeks later, Eiffel's great tower was finished. It was one of the most impressive structures built by France's "wizard of iron." Soon it would become a symbol for the "City of Light," known the world over.

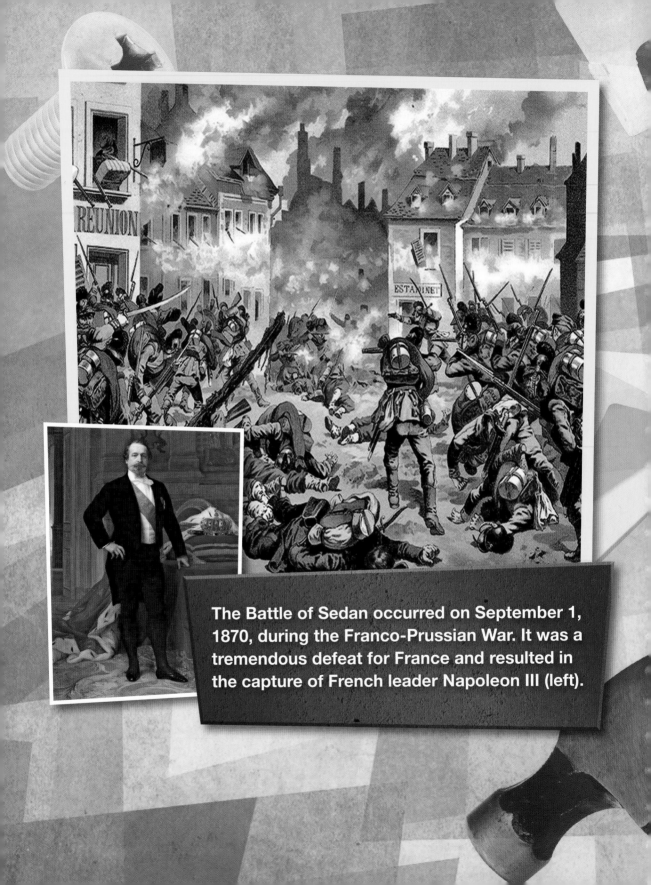

The Battle of Sedan occurred on September 1, 1870, during the Franco-Prussian War. It was a tremendous defeat for France and resulted in the capture of French leader Napoleon III (left).

The Idea

Paris lay in ruins.

The great city had been destroyed by the War of 1870. French Emperor Louis Napoleon III surrendered to the Prussians of North Germany on September 2, 1870. But a new government in Paris fought on. The Germans took over the city, and very quickly its food ran out. By November, starving Parisians were eating rats, pigeons, dogs, cats, and horses . . . even zoo animals like camels and elephants.

Peace was finally declared in February 1871, but it didn't last long. A civil battle erupted in Paris between the French army and French rebels. Both groups wanted control of the government. After months of fighting, the regular army won. By then, almost 20,000 more Parisians had died. The city was a smoking ruin.[1]

The rest of the country wasn't in much better shape. France was forced to give Germany two of its eastern provinces—Alsace and Lorraine—and to pay them five billion francs (about one billion U.S. dollars). German troops would stay in France until the money was paid.

It looked as if France would be crippled for decades. But the government managed to pay the money quickly, and German troops left in 1873. The entire French economy shifted into high gear. Factory production and foreign trade boomed, and the national income soared. France recovered with amazing speed. Once again, the country was an economic powerhouse.

This was the Belle Époque (Beautiful Age) in French history. The French decided to celebrate. On November 8, 1884, French President Jules Grévy signed a decree, stating: "A Universal Exposition of the

France is one of Europe's most important countries.

Products of Industry shall be opened in Paris May 5, 1889, and closed October 31 following."[2]

Such an exhibition would allow France to show how far it had come since the war. Grévy chose 1889 because it marked the 100th anniversary of the 1789 French Revolution.

Édouard Lockroy, the French minister of commerce and industry, had an idea: Why not build a 1,000-foot-tall tower as an unforgettable symbol of the fair? It

Édouard Lockroy

would be, he said, "a pole around which humanity spirals eternally upward."[3]

On May 2, 1886, Lockroy asked for bids and designs for the project. With a May 18 deadline, those interested in building such a tower had only 16 days to design it. More than 100 designs were submitted.[4]

One looked like a giant sprinkler. It would be able to water Paris in case of a drought. Another tower had a giant electric light-and-mirror system at the top that would light up the whole city.

But another design was ultimately chosen. It had been sketched by engineer Maurice Koechlin on June 6, 1884. The idea was developed by engineer Émile Nouguier and architect Stephen Sauvestre. However, it was another name attached to the project that brought it attention. That name was Gustave Eiffel.

Industrial exhibitions have been held around the world about every five years since 1851. Also called the World's Fair, these exhibitions became popular as a way for nations to display their progress. The Eiffel Tower was designed as a temporary structure to serve as the entrance arch for the World's Fair in Paris in 1889. The plan was to demolish it in 1909, but it was saved after the tower was turned into a giant radio antenna.

Gustave Eiffel wanted to study at the world-famous École Polytechnique, but he was not accepted.

The Builder

Alexandre-Gustave Eiffel was born on December 15, 1832, in Dijon, France. His father was a soldier named François-Alexandre. His mother was Mélanie Moneuse. His parents worked constantly, so Gustave lived with his blind grandmother. Even so, he stayed close with his parents. He often confided in his mother, and every Sunday he walked with his father. He found school boring and thought it was a waste of time. He must have learned something, though, because he graduated with degrees in both the humanities and science.

In 1850 he went to Paris to study for the entrance exams for the famous École Polytechnique (technical college). Still not happy about going to school, he wrote to his mother: "I am sad . . . we have nothing to do . . . I am bored. . . ."[1] The Polytechnique did not accept him.

Instead he enrolled in l'École Centrale des Arts et Manufactures. His uncle was a chemist, so Gustave studied chemistry. His plan was to return to Dijon to work in his uncle's factory. Unfortunately, in 1855 the uncle and Gustave's family argued. The job offer was off the table.

Forced to find another career, Eiffel took a job in 1856 with Charles Nepveu, a railway construction engineer who ran a business in Paris. At a salary of about $30 per month, Eiffel became Nepveu's private secretary. His job was to study, as he put it, "the questions which M. Nepveu was preoccupied with, notably that of foundations in rivers."[2]

By helping his boss, Eiffel learned important engineering techniques. Unfortunately, the arrangement lasted less than a year because Nepveu's company went bankrupt. However, the two men continued to work together.

In 1858 Eiffel began building a 1,670-foot bridge over the Garonne River near the city of Bordeaux. The bridge would join two rail lines. Once finished, it would be the longest bridge in Europe.

The foundation had to be strong enough to withstand the river's swift and unsettled waters. To increase its strength, Eiffel developed a new method of pile driving. A pile driver is a mechanical device that hammers poles deep into soft ground. Eiffel used compressed air and hydraulics to drive the foundation materials deep into the riverbed. He was becoming known as a talented and inventive engineer.

Eiffel's personal and business lives were both thriving. In 1862 he married Marie Gaudelet. In 1866 he opened an engineering company, Compagnie des Établissements Eiffel. He listed his specialties as "iron constructions, market halls and builders' hardware, reservoirs, gasholders, boilers, and in general all metal constructions."[3]

One of his first projects was building a gigantic exhibit hall for the 1867 Universal Exhibition in Paris. He found that cast iron has elastic properties—when it is bent, or deformed, it will return to its original shape when the deforming force is removed.

Another important project was the great Palais Des Machines built for the 1878 Paris Universal Exposition. Eiffel used wrought iron arch girders and curved support beams. These supports were lightweight and attractive yet incredibly strong. His unique use of wrought iron and the light, airy look of these structures became his trademark.

Over the years, Eiffel's company built over 40 railway bridges and viaducts. He also built the Budapest Western Railway Station in Hungary and designed the movable dome

Palais Des Machines

Budapest Western Railway Station

for the Nice Observatory in France.

By 1886, Gustave Eiffel was France's most celebrated builder. That May, when Édouard Lockroy asked for bids for a tower for the Industrial Exhibition, he already had Eiffel in mind. In fact, several years earlier, Eiffel had told Lockroy his idea for that type of tower. Lockroy knew Eiffel would be able to build it.

On January 8, 1887, Lockroy, the City of Paris, and Eiffel signed a contract. Eiffel's company would front $1.3 million of the estimated cost. The company would then receive $300,000 in each of three installments, or $900,000. For the rest of the cost, the contract gave Eiffel the right to operate the tower. He could also open restaurants, cafés, and other businesses on it for 20 years. After that, Paris would own the tower and its businesses. With the contract signed and the money in place, it was time for Eiffel to begin building the tower that would bear his name.

Gustave Eiffel was also the engineer who designed the wrought iron skeleton for the Statue of Liberty. Located on Liberty Island in New York Harbor, the colossal Lady Liberty was given as a gift from France to America in 1886.

Workers pump water out and pump air into the giant can-like metal caissons.

The Foundation and Tower

Eiffel had just two years to build the tower, so he started right away. First he took soil samples from the Champ de Mars, where the tower would stand. Beneath much of the site, dense clay soil 50 feet deep rested on a solid chalk foundation. However, closer to the nearby Seine River, the upper soil layer was loose sand and gravel. This meant that two of the tower's legs would be on firm material, while the two legs near the river would be on softer earth.

Eiffel dug deeper and finally found solid clay soil all around. He would make the foundations for the legs in the looser soil 16 feet deeper than the foundations for the other two legs.

Digging began on the four foundations on January 28, 1887. To keep water from the Seine from seeping into the holes, Eiffel used something he was familiar with from bridge building: sheet metal caissons, 16 in all (four for each side). Each of the rectangular caissons was 50 feet long, 20 feet wide, and 10 feet deep. Each weighed 34 tons. Using the caissons was like inserting giant metal cans into the earth.

Several men entered each caisson through an airlock on top. Working inside, the men used shovels and pickaxes to loosen soil and place it into buckets for removal. People above hauled up the buckets of soil and dumped them.

When 40,500 cubic yards of earth had been removed from the four sites, it was time to set the foundations. Workmen poured quick-setting cement 20 feet deep into each foundation hole. On top of this were

The mountings could be adjusted for the proper angle.

placed mammoth blocks of limestone, then two thick layers of cut stone.

Two gigantic anchor bolts, each 26 feet long and 4 inches in diameter, were embedded in the stone. They were attached to a metal base called a shoe. The tower's weight would keep it from tipping over, so the bolts were not really necessary—but Eiffel wanted to be sure that the tower was secure.

Now Eiffel became creative. He placed a piston in the hollow section of each shoe. These pistons, each of which could lift 900 tons, acted as hydraulic jacks. If the platform was the slightest bit uneven, the tower would lean as it rose upward.

The foundations were finished by the end of June 1887. The piers were set. It was time to begin building the tower.

Eiffel debated between using steel and iron for the tower. He finally chose iron because of its greater elasticity—its ability to bend and not break in high winds.

The tower was one of the first examples of large-scale industrial construction. Every calculation was extremely precise. Eiffel's draftsmen drew blueprints for every piece of metal used for the tower. There were 5,329 drawings to show 18,038 different pieces of wrought iron.[1] The drawings took 30 draftsmen 18 months to create. They covered 14,352 square feet of paper.[2]

Because the drawings were so precise, when the tower pieces arrived from the factory, all the rivet holes had already been punched in just the right place. The workers only had to line up the pieces and connect them. Many of the pieces were held together with bolts until the rivets could be placed. Some 250 to 300 workers would eventually use 2.5 million rivets to put the tower together.

Another cutting-edge idea was Eiffel's use of creeper cranes. These 13-ton cranes were attached to the tower. They ran on the tracks that

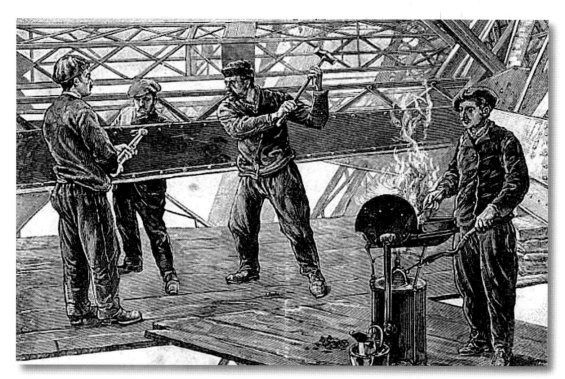

A team of four men was needed for each rivet used: one to heat it, another to hold it in place, a third to shape the head, and a fourth to install it with a sledgehammer.

Chapter 4

Creeper cranes helped transport materials up and down the tower.

would later be used for the elevators. As work progressed, the cranes moved higher and higher as needed. They made it much easier to bring materials up and down the tower.

Many people came to the construction site to watch the creeper cranes. "It is curious and interesting to watch these four lifting machines climbing ever higher along with the iron columns," one reporter wrote.[3]

Eiffel understood that a structure survived strong winds not by being super heavy, but by giving the wind as little to push and tear apart as possible. Most of the tower allows the wind to blow right through it. There is little extra metal in the tower's light and airy design. If the tower's girders and braces were melted down and spread out in an area equal to the area of its base, the sheet would be just two and a half inches thick.[4]

Even when the tower was in the design phase, Eiffel had been planning for the wind. The curved edges of its piers would reduce the wind's power. "[T]he uprights appear to burst out of the ground, and in a way to be shaped by the action of the wind," he had written.[5]

Steadily the tower rose. When the first level was completed on April 1, 1888, the tower was 180 feet tall. By August 14 of that year, with the second level finished, it was 380 feet high. It was 645 feet high in mid-December. By the end of March 1889, it had reached its final height of 984 feet. With the addition of a flagpole, it reached 1,000 feet high. To go from the bottom to the top, a person must climb 1,665 steps. Now that's a tall tower!

Eiffel painted his tower reddish brown. The bottom was darker, and the tower gradually got lighter toward the top until it was almost yellow. The project was essentially finished except for the elevators. The completed structure weighed 10,000 tons and was lit by 10,000 gas streetlamps. (Today it is lighted by 20,000 lightbulbs.)

Now, the big question was: Would people come to see it?

The tower as it stood newly finished in 1889. It takes 50 to 60 tons of paint to cover the Eiffel Tower. It is repainted every seven years— completely by hand.

A poster advertising the exposition was dominated by the spectacular Eiffel Tower—as was the exposition itself.

The Universal Exposition

All along, when it came to building the Eiffel Tower, there were people who said it couldn't be done. There were also people who didn't want it to be done. Many writers, painters, and other artists feared that the tower would destroy the beauty of Paris. Among them were the famous writers Alexandre Dumas and Guy de Maupassant.

On February 14, 1887, these protestors wrote a heated letter to France's minister of public works. In it they called the tower "a black and gigantic factory chimney" and an "odious column of bolted metal."[1]

In October that year, one of the tower's neighbors sued to stop construction. The project was restarted only when Eiffel agreed to pay for any accident that might happen.

People continued to predict disaster for the tower. One mathematics professor claimed that the tower would collapse when it reached 748 feet high. In early 1888, a headline in the daily newspaper *Le Matin* screamed, "The Tower Is Sinking!" The writer believed the tower should be demolished immediately.

But the tower neither collapsed nor was demolished. The Universal Exposition in Paris opened on May 6, 1889. On May 15, the tower was open to the public. After two years, two months, and five days of work, Eiffel was able to write: "At last!"[2]

Even though the elevators were still not ready, nearly 30,000 people paid forty cents for a ticket to climb to the first platform. More than 17,000 paid sixty cents for a ticket to climb to the second. The first elevator began operating on May 26, followed by a second on June 16. A ticket for a ride to the top cost one dollar.[3]

Even though the elevators were now working, some still bravely walked back down the stairs from the top. One man said he felt "like an ant coming down the rigging of a man-of-war [a large ship]."[4]

The tower was a runaway hit. From May 15 to November 6, nearly two million people visited the tower. That is an average of nearly 12,000 per day. One day—June tenth—23,202 visitors came to see it.[5]

Eiffel became a wealthy man from the entrance fees, the profits from tower restaurants, and the sale of souvenirs. He received much praise for the tower that many had ridiculed.

Several years later, Eiffel closed his business. He never built another major structure. His wife had died in 1877, and he lived the rest of his life in an apartment in the tower. From there he experimented in meteorology (weather), aviation (flying), aerodynamics (how air moves), and telecommunications (telegraph, radio, and telephone). He died on December 27, 1923, at age 91.

About 7 million people climb the Eiffel Tower every year, and since its opening in 1889, it has had almost 250 million visitors. It has become the most important item in the Paris skyline and one of the best-known structures in the world.

Gustave Eiffel, 1888

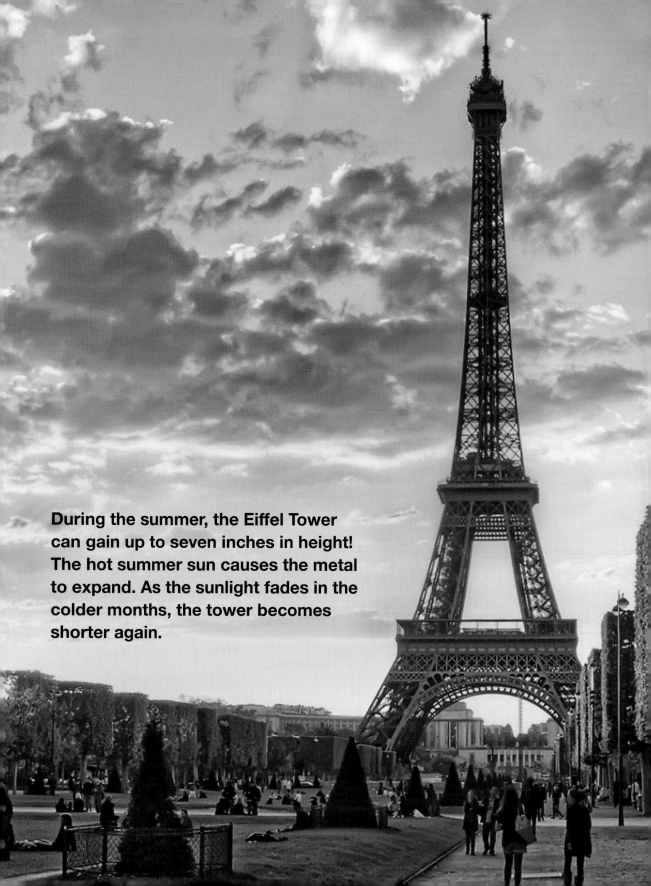

During the summer, the Eiffel Tower can gain up to seven inches in height! The hot summer sun causes the metal to expand. As the sunlight fades in the colder months, the tower becomes shorter again.

1832   Alexandre-Gustave Eiffel is born on December 15 in Dijon, France.

1855   Eiffel graduates from l'École Centrale des Arts et Manufactures (College of Art and Manufacturing) in Paris. After graduation, he specializes in metal construction, most notably bridges.

1858   Eiffel begins using a hydraulic and compressed air system of pile driving.

1884   French President Jules Grévy signs a decree stating that an Industrial Exhibition shall be held in Paris in 1889, on the 100th anniversary of the French Revolution.

1886   The Centennial Exposition Committee invites French architects and engineers to submit building designs for a 1,000-foot-tall tower. The Eiffel Tower is declared the winning design.

1887   Work on the tower's foundation begins on January 28.

1888   The first floor is finished on April 1.

1889   The Universal Exhibition in Paris opens on May 6. The Eiffel Tower is open to the public on May 15.

1909   A permanent underground radio center is built near the south pillar of the tower, saving the tower from destruction.

1923   Gustave Eiffel dies on December 27 at age 91.

The top of the Eiffel Tower

Chapter Notes

Chapter 1. The Terrifying Climb
1. Jill Jonnes, *Eiffel's Tower* (New York: Viking, 2009), p. 66.
2. Ibid.
3. Ibid., p. 67.
4. Ibid., p. 68.
5. Ibid.

Chapter 2. The Idea
1. Joseph Harriss, *The Tallest Tower* (Boston: Houghton Mifflin, 1975), p. 4.
2. Ibid., p. 6.
3. Frederick Brown, *For the Soul of France* (New York: Alfed A. Knopf, 2010), p. 125.
4. Harriss, p. 11.

Chapter 3. The Builder
1. Henri Loyrette, *Gustave Eiffel* (New York: Rizzoli International Publications, 1985), p. 29.
2. Ibid., p. 30.
3. Ibid., p. 37.

Chapter 4. The Foundation and Tower
1. Harriss, Joseph, The Tallest Tower. (Boston: Houghton Mifflin Company, 1975), p. 61.
2. Joseph Harriss, *The Tallest Tower* (Boston: Houghton Mifflin, 1975), p. 61.
3. Ibid., p. 67.
4. Ibid., 60.
5. Frederick Brown, *For the Soul of France* (New York: Alfed A. Knopf, 2010), p. 140.

Chapter 5. The Universal Exposition
1. Jill Jonnes, *Eiffel's Tower* (New York: Viking, 2009), p. 163.
2. Joseph Harriss, *The Tallest Tower* (Boston: Houghton Mifflin, 1975), p. 116.
3. Ibid., p. 116.
4. Mary McAuliffe, *Dawn of the Belle Epoque* (New York: Rowman & Littlefield Publishers, 2011), p. 196.
5. Henri Loyrette, *Gustave Eiffel* (New York: Rizzoli International Publications, 1985), p. 162.

Books

Brown, Jeff. *Flat Stanley's Worldwide Adventures #11: Framed In France.* New York: HarperCollins, 2014.

Holub, Joan. *What Is the Statue of Liberty?* New York: Grosset & Dunlap, DGS edition, 2014.

Pizzoli, Greg. *Tricky Vic: The Impossibly True Story of the Man Who Sold the Eiffel Tower.* New York: Viking Books for Young Readers, 2015.

Stevenson, Steve. *The Eiffel Tower Incident (Agatha: Girl of Mystery).* St. Louis, MO: Turtleback Books, 2014.

Wilbur, Helen L. *E Is For Eiffel Tower.* Ann Arbor, MI: Sleeping Bear Press, 2010.

On the Internet

Discover France: "The Eiffel Tower, Paris"
http://discoverfrance.net/France/Paris/Monuments-Paris/Eiffel.shtml

Eiffel Tower Guide
http://eiffeltowerguide.com/

Eiffel Tower: Official Web Site
http://www.toureiffel.paris/en

History Channel: Eiffel Tower Fact & Summary
http://www.history.com/topics/eiffel-tower

Works Consulted

Brown, Frederick. *For the Soul of France.* New York: Alfred A. Knopf, 2010.

Harriss, Joseph. *The Tallest Tower: Eiffel And The Belle Epoque.* Boston: Houghton Mifflin Company, 1975.

Horne, Alistar. *Seven Ages of Paris.* New York: Alfred A. Knopf, 2004.

Jonnes, Jill. *Eiffel's Tower.* New York: Viking, 2009.

Loyrette, Henri. *Gustave Eiffel.* New York: Rizzoli International Publications, Inc., 1985.

McAuliffe, Mary. *Dawn of the Belle Epoque: The Paris of Monet, Zola, Bernhardt, Eiffel, Debussy, Clemenceau, and Their Friends.* New York: Rowman & Littlefield Publishers, Inc., 2011.

Palermo, Elizabeth. "Eiffel Tower: Information and Facts." *Live Science*, May 7, 2013. http://www.livescience.com/29391-eiffel-tower.html

Salvadori, Mario. *Why Buildings Stand Up.* New York: W.W. Norton & Company, 1980.

aerodynamics (ayr-oh-dy-NAM-iks)—The study of the motion of air and other gases.

antenna (an-TEN-uh)—A wire that receives radio and TV signals.

aviation (ay-vee-AY-shun)—The business of flying airplanes and other aircraft.

bankrupt (BANK-rupt)—To owe more money than one is taking in, allowing creditors to take ownership of one's property.

blueprint (BLOO-print)—A detailed drawing of a construction project.

caisson (KAY-sun)—A watertight chamber used in construction work under water or as a foundation.

canal (kah-NAAL)—A waterway dug to allow boats or ships to pass through.

commerce (KAH-murs)—Trade; the buying and selling of goods.

elasticity (ee-las-TIH-sih-tee)—The ability to bend or stretch.

exhibition (ek-sih-BIH-shun)—A public display.

girder (GIR-der)—A a strong beam used to build buildings.

humanities (hyoo-MAN-ih-tees)—The study of philosophy, arts, or languages.

hydraulic (hy-DRAW-lik)—Operated or moved by liquid, usually water or oil.

illuminate (ih-LOO-muh-nayt)—To light; to shed light.

innovative (IN-oh-vay-tiv)—Introducing something new or different.

inquiry (IN-kwuh-ree)—An official examination in order to gain information.

lock—On a canal, a section with gates at each end that allow water in or out in order to raise or lower ships or boats to the next level.

meteorology (mee-tee-or-AH-luh-jee)—The study of weather.

piston (PIS-tin)—A disk that moves back and forth inside a cylinder.

rigging (RIG-ing)—The ropes used to support the sails of a ship.

rivet (RIH-vit)—A metal pin with a head on one end that is used to join two pieces of metal; the shank that passes through the metal pieces is hammered to create a second head.

siege (SEEDJ)—In wartime, to surround an enemy area so that it cannot get help or supplies.

telecommunications (teh-leh-kuh-myoo-nih-KAY-shuns)—The science of sending messages over long distances by telephone, satellite, radio, and other electronic means.

trademark (TRAYD-mark)—A feature, name, or logo (symbol) that identifies a person or company.

viaduct (VY-uh-dukt)—An elevated roadway usually supported by arches, pillars, or piers.

PHOTO CREDITS: p. 1—Nicki Dugan Pogue; pp. 4, 6, 7, 8, 10, 14, 16, 20, 21—Public Domain; p. 19—Brodi Broda; p. 21—Nga.gov. All other photos—Creative Commons. Every measure has been taken to find all copyright holders of material used in this book. In the event any mistakes or omissions have happened within, attempts to correct them will be made in future editions of the book.

Index